Earth: The Water Planet

by E.C. Hill

Table of Contents

Introduction

All living things need water to live. That includes you. Your body is made up of about 65% water. Water makes life possible.

We can find water in many places on Earth. When astronauts look at Earth from space, what they see is mostly water. One astronaut said Earth looked like "a sparkling blue and white jewel." The blue he saw was the blue of the oceans. The white was the clouds in the sky.

▲ This watering hole is in Etosha National Park, Namibia.

Read this book to learn more about water and its role in Earth's environment. Find out about the great oceans. Learn how they affect the world's **climate**. Then learn where fresh water comes from and how we use it.

Oceans

Water covers much of Earth. Most of that water is in the oceans. There are five major oceans. They are the Pacific, Atlantic, Indian, Southern, and Arctic oceans. How are the oceans alike? Well, one thing that they all share is their **salinity**. Water from any ocean is 3.5% salt and 96.5% water. You can taste the saltiness in ocean water.

They Made a Difference

Otis Barton and William Beebe were the first people to go down to the deepest parts of the ocean. The two men designed the first bathysphere to take them into the deep sea. It looked like a steel ball with a window in it. The two men used it to dive over 3,000 feet (914 meters) down into the ocean. A 400-pound (181-kilogram) door locked it tight.

The Pacific Ocean is approximately what fraction of the world's oceans? Hint: This is a two-step problem. First you have to add together all the oceans. You can estimate to get an approximate answer.

▲ Over 70% of Earth's surface is covered by water.

Oceans

Ocean	Size
Pacific Ocean	60,060,893 sq mi (155,557,000 sq km)
Atlantic Ocean	29,637,974 sq mi (76,762,000 sq km)
Indian Ocean	26,469,620 sq mi (68,556,000 sq km)
Southern Ocean	7,848,299 sq mi (20,327,000 sq km)
Arctic Ocean	5,427,052 sq mi (14,056,000 sq km)

The Ocean Floor

The ocean floor has the same features that land does. There are mountains, valleys, hills, and plains along the ocean floor. We can't see them because they are hidden beneath the water.

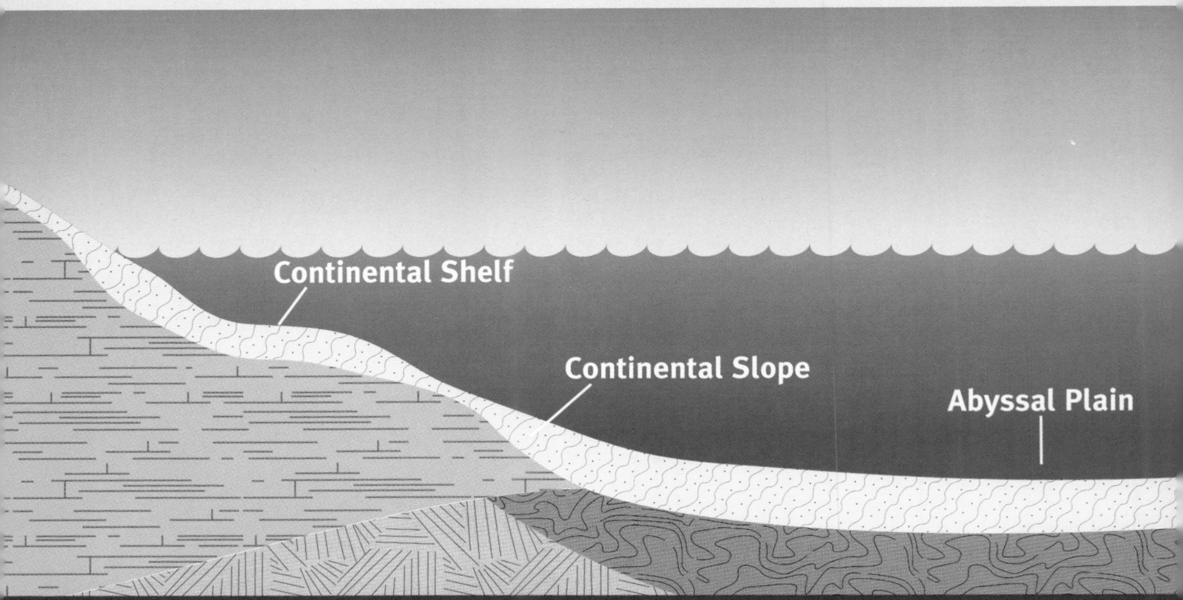

Continental Shelf

Continental Slope

Abyssal Plain

▲ The ocean floor

It's a Fact

The highest point on Earth is Mount Everest. The lowest point in the oceans is in the Marianas Trench in the Pacific Ocean. The longest mountain range is in the ocean. It's called the mid-ocean ridge. It runs for more than 50,000 miles (80,467.2 kilometers). Scientists now know that the mid-ocean ridge is part of every ocean on Earth.

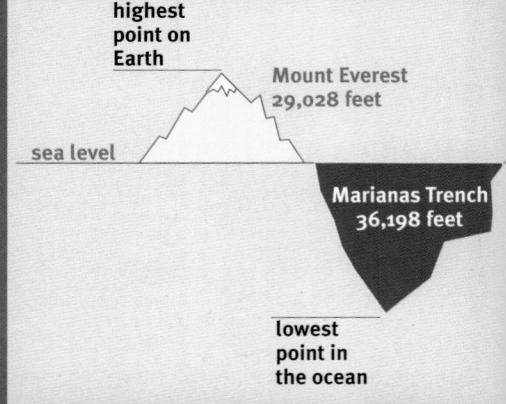

highest point on Earth

Mount Everest 29,028 feet

sea level

Marianas Trench 36,198 feet

lowest point in the ocean

There are three main areas of the ocean floor. Around the edges of the continents, the ocean gets deeper very slowly. This area is actually a part of a continent, but it is under water. It is called the **continental shelf**.

The continental shelf ends suddenly. At its edge the ocean floor drops sharply. This steep slope is called the **continental slope**. Here the ocean floor becomes deep very quickly.

At the end of the continental slope, the ocean floor gradually flattens. This flat part of the ocean floor is called the **abyssal plain** (uh-BIH-sul PLANE). The abyssal plain covers almost one-half of Earth's surface.

▲ Have you ever been to an ocean beach? When you walked into the water, you were on the continental shelf.

Oceans on the Move

Have you ever been knocked down by a wave at the beach? Did you wonder where the waves came from?

The waters of the oceans are always moving. But what makes the ocean move? The sun is the main cause. It affects the ocean in many ways. It heats the air over the ocean. Warm air is lighter than cold air. So, the warm air rises. Cooler air flows in to replace it. This makes wind. And wind blowing on the surface of the ocean pushes the water into waves. The harder the wind blows, the larger the wave.

▲ Waves moving toward the shore must travel uphill. (Remember the continental shelf.) That's why the waves slow down. But faster waves are right behind them. The water piles up and crashes against the shore.

Everyday Science

Ocean water contains many minerals. But the main mineral is sodium chloride. That's the same salt people put on food.

Currents

Wind currents blow over the ocean day after day. During the day, wind blows from the water to land. During the night, wind blows from the land to water. This makes the oceans move in strong currents. An ocean current is a large mass of water that flows like a river through the ocean.

Currents flow in all the oceans. They carry cold water to warm regions and warm water to cold regions. Ocean currents tend to move in circles.

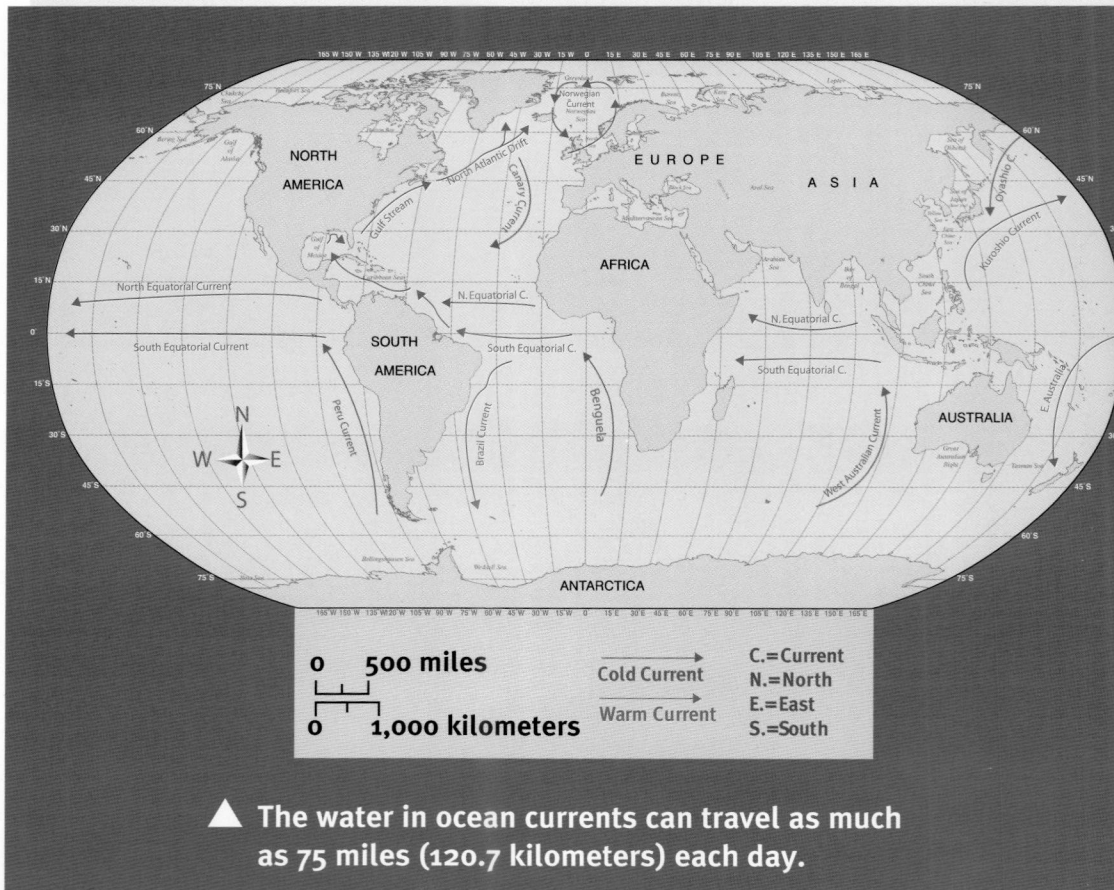

▲ The water in ocean currents can travel as much as 75 miles (120.7 kilometers) each day.

One of the major ocean currents is the Gulf Stream. The Gulf Stream flows from the warm waters near Florida. The current moves north and then east across the Atlantic. It brings warm water to northwestern Europe. The water warms the region's climate. Northwestern Europe would be colder without the Gulf Stream.

It's a Fact

Benjamin Franklin discovered that the fastest way to sail from America to England was to use the Gulf Stream. His discovery sped up the delivery of mail between America and Great Britain.

▲ Palm trees grow along the southwest coast of Great Britain. The Gulf Stream warms this area enough for palm trees to grow here.

Weather and the Water Cycle

You get out of a pool on a hot day. As you walk to your towel, you see your wet footprints. But when you turn back, they are gone. Where did they go? What happened to the water?

The heat caused the water to **evaporate**. When liquids evaporate, they turn into **vapor**, or gas. The water from wet footprints evaporates and floats in the air.

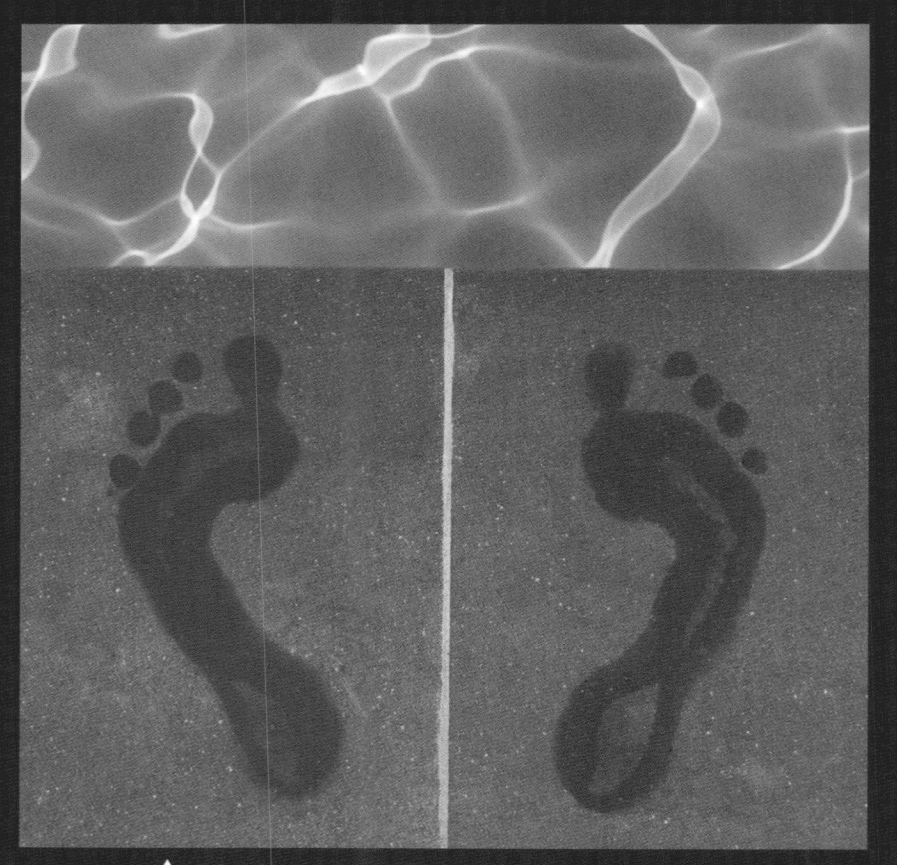

▲ The hot sun will make these wet footprints evaporate quickly.

The same thing happens with oceans and other bodies of water. The sun heats the water. Then huge amounts of water evaporate. The water vapor rises into the air. As the water vapor rises, it cools and **condenses**, or turns from a gas into a liquid. Droplets of water form clouds.

When the droplets get too heavy to float in the clouds, rain falls. If rain were spread evenly across the surface of Earth, about 40 inches (101.6 centimeters) of rain would fall each year.

The Water Cycle

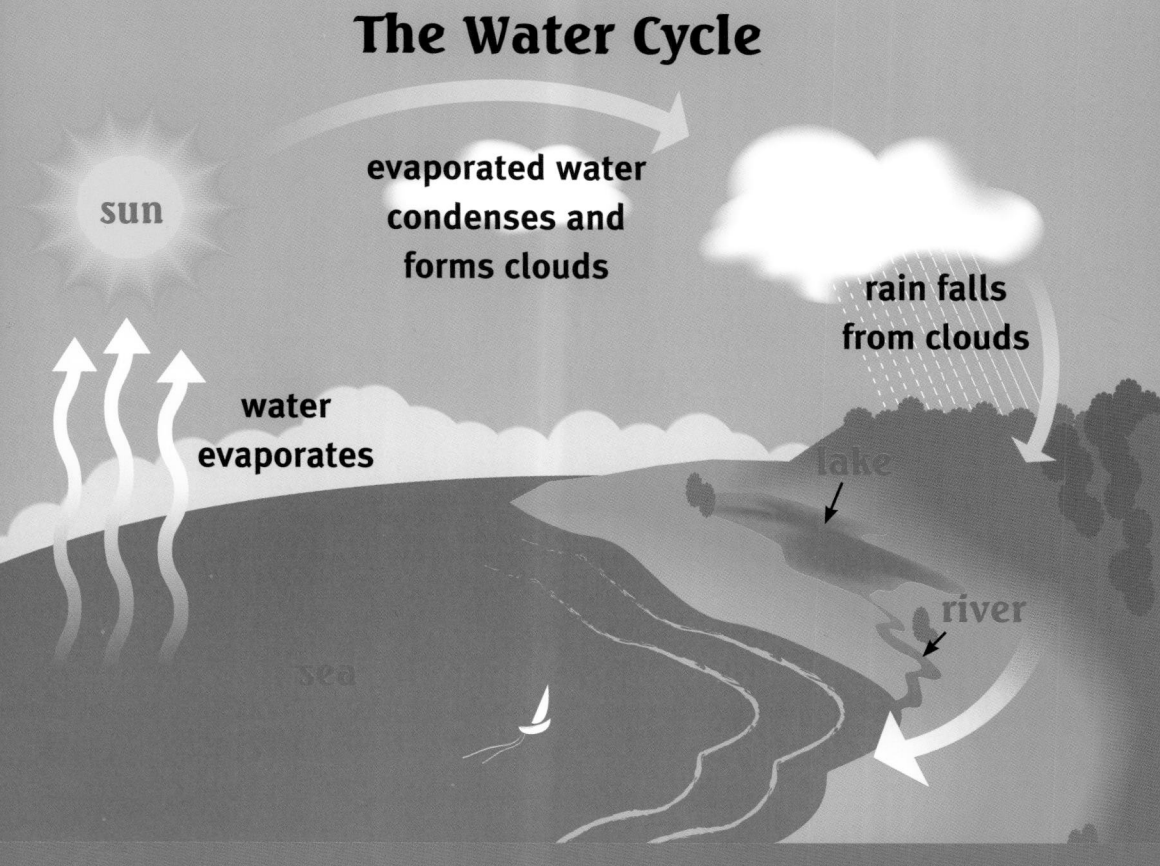

sun

evaporated water
condenses and
forms clouds

rain falls
from clouds

water
evaporates

lake

river

sea

▲ Water evaporates from lakes, rivers, and other bodies of water. However, nearly 80% of all evaporation is from the oceans.

The water is not spread evenly, though. Rain falls heavily in some places. Other places get very little rain. Wind and rain follow patterns. One pattern is the **monsoon.**

The monsoon affects India and other parts of Asia. During the summer months, the land heats up under a strong sun. The air above the land heats up as well. The air rises. The hot air rising over the land pulls very moist air in from the Indian Ocean. Heavy rain then falls in parts of India.

▲ Some of the wettest places on Earth are in India. Some areas get over 400 inches (1,016 centimeters) of rain in an average year.

Clouds and the Wind

Have you ever watched clouds blow across the sky? You were watching water move. Wind moves the clouds. And the sky is in constant motion, just like the sea. So clouds are almost always moving in the sky.

But not all the water stays in the clouds as droplets. Some falls as rain, snow, sleet, or hail.

▲ Clouds cover about 60% of the sky at any time.

✔ Point

Make Connections

What other fiction or nonfiction books have you read about clouds? How was the information the same as or different from what you read in this book?

Careers in Science

Do you wonder why different locations have different climates? Then being a climatologist might be the job for you. Climatologists are scientists who study the average weather conditions over a long period of time. One of their jobs is to gather information about the causes of weather patterns. Climatologists are worried that Earth's climate is changing in dramatic ways.

Completing the Water Cycle

The energy of the sun is one of the great forces that moves water around the planet. Another force is **gravity**. Gravity is the force of attraction between Earth and all objects on Earth. When you drop a ball, gravity makes it fall to the floor. Gravity also makes rain fall to the ground.

It's a Fact

The amount of water in the planet's water cycle never changes. The water that falls as rain or snow has existed for billions of years. It has moved through the water cycle countless times.

▲ Storms, like hurricanes, carry large amounts of water from the sea to the land.

Most rain sinks into the ground. But sometimes there is too much rain. The rain that doesn't sink into the ground becomes **runoff**. Runoff is water that flows over the surface of land into streams and rivers. Streams and rivers always flow downhill because of gravity. The water in rivers and streams keeps moving until it reaches the lowest point it can. That lowest point is often sea level. When water reaches the ocean, it completes the water cycle.

✔ **Point**

Reread
Reread page 13 to find out where the water in clouds comes from.

▲ **Salak River in Borneo**

Fresh Water

When water evaporates, the water vapor is pure. If it comes from the sea, none of the sea's salt evaporates with it. For this reason, water from rain is fresh water. Where else can you find fresh water?

Fresh water is in lakes. Lakes are bodies of water that are surrounded by land. Ponds are much like lakes, but smaller.

▲ There are millions of lakes in the world. Canada alone has about two million.

Water from streams and rivers often flows into lakes. Water also passes into lakes from **springs**. Springs are places where underground water comes to the surface.

Lake Superior is the largest freshwater lake in the world. But another lake holds even more fresh water than Lake Superior. Lake Baikal in Asia holds more water because it is so deep. It is the deepest lake in the world.

Lake	Depth
Superior	1,333 feet (406 meters)
Baikal	5,370 feet (1,637 meters)

▲ Lake Superior is the deepest of the Great Lakes at over 1,300 feet (396.2 meters) deep. Lake Baikal is about four times as deep.

It's a Fact

Every living thing needs water to survive. Most animals are mostly made of water. Plants also contain plenty of water. Even the driest plants are about 50% water.

Rivers

Rivers are different from lakes. Lakes hold water, but water flows through rivers. River water may move quickly or slowly, but it moves all the time.

Many rivers start high in the mountains. Mountain streams rush downhill from the high ground. The flow of water slows when the river reaches flatter ground.

▲ Yangtze River

2. Solve This

How much longer is the Nile River than the Yangtze River?

The World's Longest Rivers

River	Length
Nile (Africa)	4,100 miles (6,598.3 km)
Amazon (South America)	4,000 miles (6,437.4 km)
Mississippi/Missouri (North America)	3,800 miles (6,115.5 km)
Yangtze (Asia)	3,700 miles (5,954.6 km)
Yenisei (Asia)	3,400 miles (5,471.8 km)

Even the largest rivers start small. Let's go to South America where the mighty Amazon River begins. It starts as a small trickle of water by a cliff in the Andes Mountains. The stream moves quickly downhill. It grows larger as other streams join it. Over 1,000 streams and rivers add their waters to the Amazon. That is how it has become one of the longest and largest rivers in the world.

✔ Point

Visualize

Use the information on this page to draw what you think the Amazon looks like at its beginning.

▲ The Amazon River holds nearly one-fifth of the river water in the world.

Underground Water and Ice Caps

Rivers and lakes are not the only bodies of fresh water. There are two other sources of fresh water. Each one holds far more water than lakes and rivers combined.

The first is underground water. As water sinks into the ground, some of it passes deep into the earth. The water soaks into the sand, gravel, or stone. It builds up over thousands of years. Huge amounts of water build up underground. These underground layers of the earth that hold water are called **aquifers** (A-kwih-ferz).

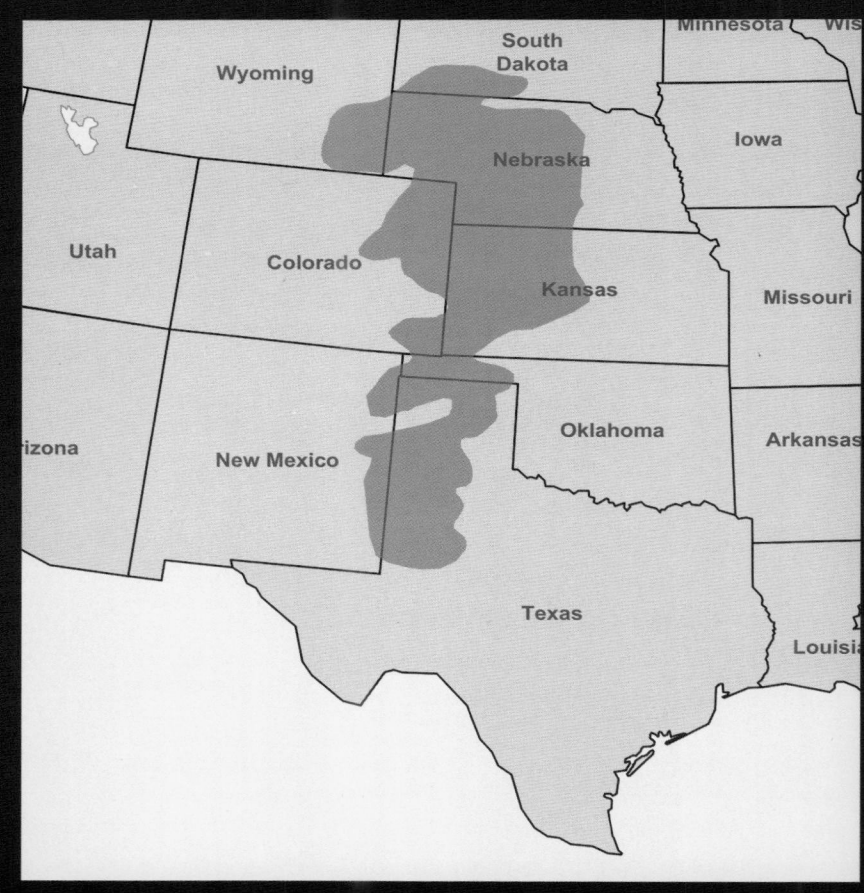

▲ The Ogallala Aquifer is an important source of underground water. It's found under the Great Plains of the United States.

Another form of fresh water is ice. Great amounts of fresh water have built up as ice in cold places. **Ice caps** cover Antarctica and the island of Greenland. Ice caps form when snow falls but does not melt fully. The snow builds up. Over time the snow becomes packed into layers of ice. The ice caps on Antarctica are nearly 3 miles (4.8 kilometers) thick in places. Some of the ice may be over 400,000 years old.

▲ Glaciers form the same way ice caps do. Many glaciers are found in high mountains, where snow builds up over time.

Freshwater 3%

Saltwater
97%

All other forms
31%

Freshwater frozen
in ice and snow
69%

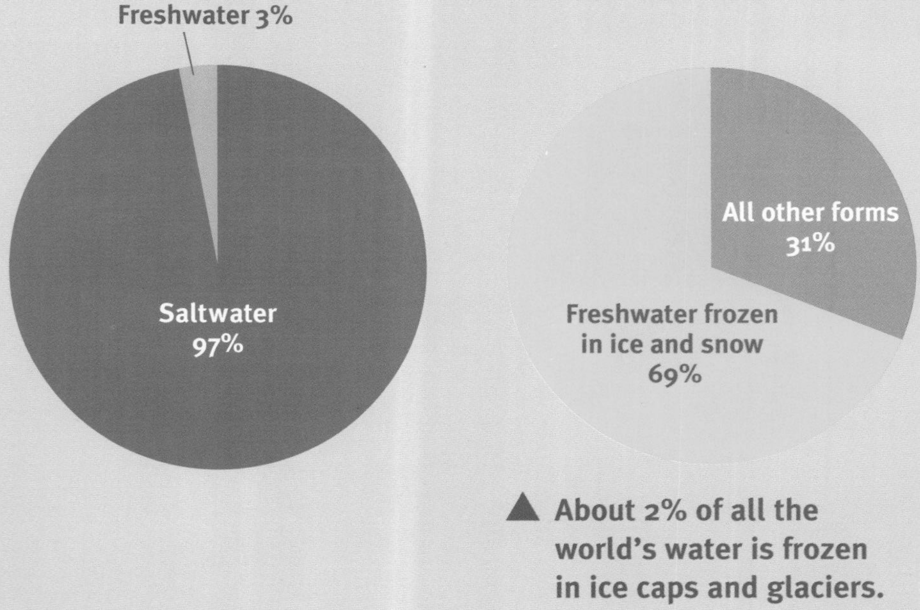

▲ About 2% of all the
world's water is frozen
in ice caps and glaciers.

Using Water

How do you use water? You drink it, you wash with it, and your family cooks with it.

You also depend on water in ways you might not think about. Farmers need water to grow the foods we eat. They give water to the cattle, pigs, chickens, and other animals they raise.

It's a Fact

Americans are lucky to have a large supply of water in most areas of the country. People in the United States use a great deal, too. For inside uses, like drinking, cooking and washing, Americans use seventy-four gallons of water per person every day.

3. Solve This

In which of the ways do Americans use the most water? Which use is just over 1/5 of the total?

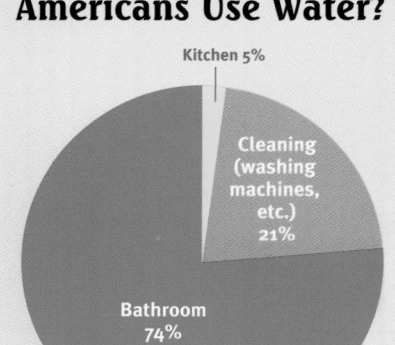

How Do Americans Use Water?

Kitchen 5%

Cleaning (washing machines, etc.) 21%

Bathroom 74%

The electricity in your home may also come from water. Dams on rivers can produce electricity. The dams have special machines built into them. The machines are called **turbines.** Turbines use the force of flowing water to make electricity.

Water has another use in people's lives. Water is fun! Millions of people around the world enjoy fishing, swimming, sailing, and canoeing in the water.

▲ Millions enjoy lakes in the summer. Some even fish on frozen lakes in the winter.

▲ Rivers flowing into Canada's Hudson Bay are dammed to produce electricity.

Overuse

People don't always use water wisely. Huge amounts of clean water are pumped into cities for people to use. After it is used, the water is dirty. Cities must find a way to get rid of the wastewater.

In the past, the wastewater passed into rivers or the sea. Today much of the water is cleaned before it is released.

◄ Israel has a dry climate. That's why it uses all the available fresh water to meet its needs.

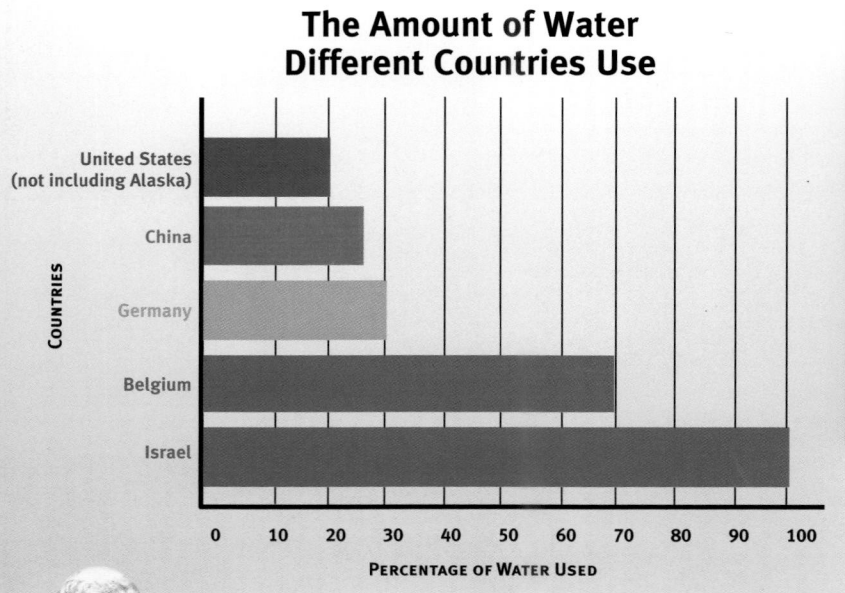

The Amount of Water Different Countries Use

PERCENTAGE OF WATER USED

COUNTRIES

Water resources can also be overused. Water from deep underground can be pumped to the surface. Farmers in the United States use this water every day. But the water is being taken from the ground more quickly than it is being replaced. Some wells are going dry because too much water is being removed.

The Aral Sea

▲ The Aral Sea in central Asia was once the fourth largest lake in the world. Much of the water in the rivers that fed the lake was taken away for farming and other purposes. The sea has shrunk and split into several smaller lakes. Scientists worry that it may disappear completely in the future.

Water and the Future

The world's population is growing. More people will need water for drinking, washing, and growing food.

But water is a limited resource. That means there is only so much to go around. What can we do to make sure we always have enough water to meet our needs?

It's a Fact

The Florida Everglades are natural wetlands. They once covered more than four million acres. Today they cover only half that much land. But people are buying back land that was turned into farms. They plan to make the land part of the Everglades again.

Wetlands are one answer. Scientists have learned that these low, marshy areas actually make water cleaner. Wetlands do this by filtering out wastes. Some towns have built artificial wetlands to treat wastewater.

Building or restoring wetlands has a bonus. Plants and animals move onto the wetlands. People can visit to enjoy nature.

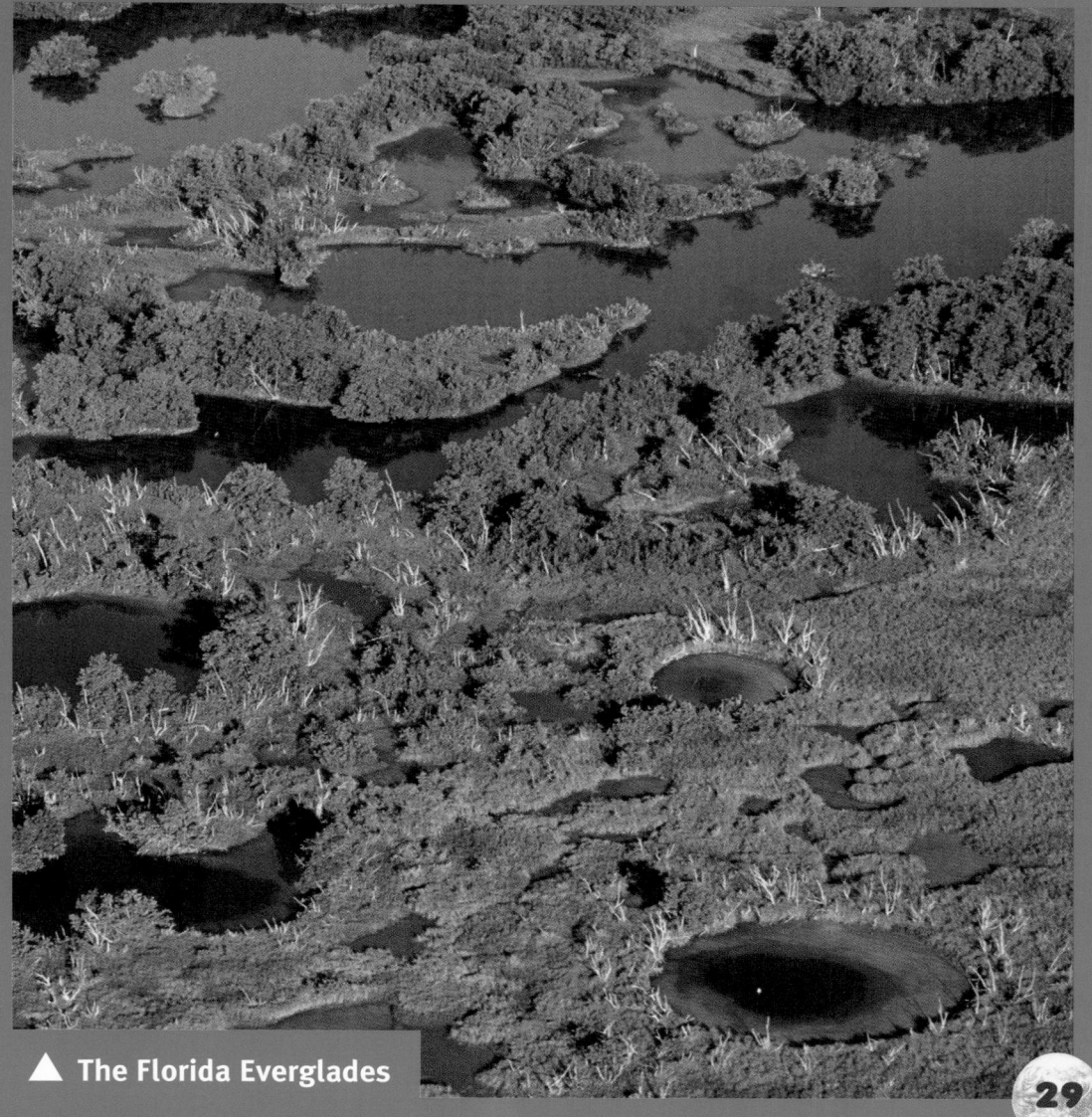

▲ The Florida Everglades

Conclusion

Earth holds a very large amount of water. Most of the water is salt water in the oceans. Water from the oceans and other sources evaporates, condenses, and forms clouds. Clouds drop rain and snow over the land.

Rain and snow produce fresh water. Fresh water can be found in lakes, rivers, underground, and frozen as ice. People use fresh water for drinking, cleaning, and raising food. We cannot live without water.

Water is a limited resource, but it also renews itself. If we learn to save water, we will always have it.

Water	
Salt water	in oceans
Fresh water	in lakes, rivers, aquifers, and ice
People use water	in many ways

Glossary

abyssal plain (uh-BIH-sul PLANE) the vast floor of the deep ocean (page 7)

aquifer (A-kwih-fer) an underground layer of sand, gravel, or stone that contains water (page 22)

climate (KLY-mut) the average weather conditions of a place or region throughout the year (page 3)

condenses (kun-DENS-is) changes from a gas to a liquid upon cooling (page 13)

continental shelf (kahn-tih-NEN-tul SHELF) the gently sloping part of a continent that is under water (page 7)

continental slope (kahn-tih-NEN-tul SLOPE) the edge of a continent that drops steeply down to the deep ocean floor (page 7)

evaporate (ih-VA-puh-rate) to become a vapor, or gas (page 12)

gravity (GRA-vih-tee) the force that pulls objects toward the center of Earth (page 16)

ice cap (ISE KAP) a thick layer of permanent ice (page 23)

monsoon (mahn-SOON) a pattern of wind and rain in the Indian Ocean and southern Asia (page 14)

runoff (RUN-auf) rain that flows over the ground and into streams (page 17)

salinity (sah-LIH-nih-tee) saltiness (page 4)

spring (SPRING) a place where underground water flows to the surface (page 19)

turbine (TER-bine) a machine that uses the power of flowing water to make electricity (page 25)

vapor (VAY-per) The gaseous state of matter (page 12)

Answers to Solve This

Page 5: about 46%   **Page 24:** bathroom; cleaning
Page 20: 400 miles (643.7 km)

Index